Audrey K. De Bouansa G.De Lumiere

INTRODUCTION TO THE TWELVE-DIMENSIONAL UNIVERSE THEORY

Audrey K. De Bouansa G.De Lumiere

INTRODUCTION TO THE TWELVE-DIMENSIONAL UNIVERSE THEORY

TO RESTORE BALANCE TO OUR WORLD

ScienciaScripts

Imprint

Cover image: www.ingimage.com

This book is a translation from the original published under ISBN 978-620-6-71151-3.

Publisher:
Sciencia Scripts
is a trademark of
Dodo Books Indian Ocean Ltd. and OmniScriptum S.R.L publishing group

120 High Road, East Finchley, London, N2 9ED, United Kingdom
Str. Armeneasca 28/1, office 1, Chisinau MD-2012, Republic of Moldova, Europe
Printed at: see last page
ISBN: 978-620-7-61718-0

AN INTRODUCTION TO THE THEORY OF THE TWELVE-DIMENSIONAL UNIVERSE TO RESTORE BALANCE TO OUR WORLD

AUDREY KIBAMBA DE BOUANSA GARE DE LUMIERE

WORK

Lightwave Research (Theoretical and High Energy Physics) with Digital and Quantum Poetry in a Twelve-Dimensional Spacetime
Digital and Quantum Epistemology specialism Science related to Spirituality and African Art course
Presented on 20 February 2024 at 10 a.m.

At the Faculty of Science and Technology of the Marien Ngouabi University in Brazzaville

By

Audrey KIBAMBA MOUNTOU

Audrey KIBAMBA DE BOUANSA GARE DE LUMIERE
Under the supervision of Professor Basile Richard BOSSOTO Dean of the Faculty of Science and Technology at Marien University NGOUABI of Brazzaville And
President of the African Mathematical Union With
The collaboration of Lauril LOMANGA OKANA

Head of the Physics Department at the Faculty of Science and Technology, Marien Ngouabi University, Brazzaville.

And

Dick Hertmann DOUMA
Head of the Master's course in Materials Science and Energy at the Faculty of Science and Technology, Marien Ngouabi University, Brazzaville.

CONTENTS

FOREWORD

In a world of complexity, uncertainty and frightening inequality, human beings are becoming increasingly disjointed and metallic, like the robots generated by artificial intelligence. Life is becoming more and more like a battle, with people preparing to parry blows, coming together, knotting up and separating at the slightest jolt. With social networks, we all reinvent what we are not. Born to be immobilised, with their eyes glued to their smartphones, human beings lose their virtue and become like placid blocks of ice clad in epidermis.

In this fascinating book, Professor Audrey KIBAMBA plunges us into a space of 12 dimensions. The aim of this space is to enable each being to both develop individually and live in solidarity with others. It is also a permanent benefit, the result of understanding others and of ourselves. The author evokes time and space, because human history does not only unfold in time, it also takes place in space. The event, on the scale of the individual, just like the slow evolution of human societies, is inscribed with the double and indissociable coordinates of time and place. As far as I'm concerned, this is a work that embraces a humanist and universalist philosophy of life, not to say a timeless one, because it brings people together without forgetting the sciences or the traditional disciplines, those 'tools' that are commonly referred to as the auxiliary sciences of history. It is a work that makes it possible to achieve what seems unattainable: the metamorphosis of the world into a more human and harmonious one. The world is not made up entirely of empty space, and time, space, matter and energy do not exist as our senses perceive them. Professor Audrey KIBAMBA stresses the challenge of teaching a way of thinking capable of connecting ancestral and contemporary knowledge, which is more than ever dispersed and compartmentalised. The famous physicist and astronomer Galileo was the first to mathematise physics; he said that the universe is a book that speaks in mathematical language. We need to prioritise knowledge for all, and we need to put an end to this world of oppression and domination that is ravaging humanity. As Edgar Morin so aptly put it: "Being The human being is at once physical, biological, psychic, cultural, social and historical". Thanks to her 12 dimensions, Professor Audrey KIBAMBA attempts to build the necessary balance between rationality and spiritual, moral and cultural values, in mutual respect between beings: an authentic life of attachment, joy, love and exaltation; a real life as opposed to current survival! The number 12 is a symbol of stability, harmony and symmetry, and in this book it conveys an undeniable sense of well-being and a human dimension so long awaited by forgotten beings...The aim of this 12-dimensional space is to solve the bewitching enigma: What is Man doing on Earth? The author proposes a true and joyful odyssey, an intellectual and incredible turning point that deserves to be tried... The researcher Audrey KIBAMBA provides answers to the most current questions about life, about Man, about this fragmented world, about values and knowledge, ethics and the sharing of wealth. In the end, the equilibrium of societies and human beings depends above all on the equilibrium of the continents...

Through this 12-dimensional space, Professor Audrey Kibamba's work calls for a profound rethinking of the human soul, and the need to centre life around the human and not around inanimate 'things'! This space allows us to leave our caves and guide us towards the path of Light. A more complete space that, through the spirit that animates it, proves to be a genuine antidote for building a world where children are born to live happily... Certain dimensions take the reader on an upward journey from the earthly to the spiritual and to the geniuses of our ancestors. This book is an invitation to work on the soul from the inside rather than the outside; it provides keys and paths to purification and inner emptying. The author's writing undeniably stirs the conscience. I would like to end this preface with a thought from Plato, considered to be the father of Western philosophy. Professor Audrey KIBAMBA's work addresses this thought with the same acuity. Plato considers that the visible world is only an imperfect copy of another world! The number of dimensions is 12, and this number contains a wealth of symbols. These symbols constitute a mysterious language that can remain obscure to those who do not possess the key. For Plato, the 12-sided regular polyhedron represents the universe and also the 12 constellations of the zodiac. Believe me, Professor Audrey KIBAMBA's book sheds light on all beings. It's an eagerly awaited book for anyone trying to understand and anticipate developments in the years ahead. Using a number of disciplines, the author sketches out the new global issues and balances, as well as their interactions. An intellectually audacious attempt to ensure that the path of life leads beings to experience the beginnings of a new humanity, both individually and collectively.

Professor Hassan Bakhsiss

INTRODUCTION

From the depths of the Algebraic and Geometric Constellation of the Great Dog, only a scientist can recognise the value of another scientist on the basis of his achievements and contributions to the progress of humanity in the universality of the light of reason.

The scientist is the person who knows, who understands and who acts in the light of the wisdom of reason based on common sense, the most widely shared principle in the world of ideas, through the intellectual coherence of mathematical thought in a scientific approach.

Research and Discovery are a shared passion in a relationship of common destiny in the light of the wisdom of love through a refinement of thought. Research is a genius of rare quality whose imagination always leads us to give birth to our emotions that speak to us in practical experience drawn from the architecture of the true truth that reflects reality itself.

Every Theory of the universe is a child of its time in the eternal memory of generations whose intelligences escape from one another in the foliage of time. There can be no Scientific Theory without Philosophy drawn from the basis of Exoteric Spirituality in know-how, which must be accompanied by know-how in order to rediscover the balances of our world in the interconnection between the wave dimension and the physical dimension.

The Twelve-Dimensional Theory of the Universe lays the new foundations for the revolution in universal mathematical thought, based on interdisciplinarity, multidisciplinarity and transdisciplinarity - essential assets for access to true knowledge, in the totality of knowledge and in the knowledge of the totality in the culture of the search for excellence in human competence.

Titles or degrees in Scientific Research are not pearls to be given to the gentlemen of science as a certificate to belong to a club of friends without the slightest achievement or contribution in the field of science and literature. You cannot be master or doctor in someone else's theory. That's not possible. A doctor is only someone who has written an original thesis with the intellectual capacity to produce new knowledge that contributes to the progress of humanity. So I whitened my nights until I transformed the darkness of the universe into the fuel of the day on a slope of sweat in the perseverance and prudence of time. Where the nights mourn like moons and the stars like totems.

Then I came to the obvious conclusion that every universe is nothing but disorder, coupled to order by the reality of space-time. This is what even justifies the fact that in every universe things are linked one after the other and in mathematical relationship with a moving earth of surveying.

Restoring balance to our world means solving the problems associated with global warming in the Theory of Disorder using the formulas of the new pure mathematics of nature.

The ineffectiveness of African universities in innovation, invention and the creation of the mind leads us to redefine the function of our universities in research and discovery that contribute to the development of nations.

In the United States, for example, it's the universities that make the discoveries. And in Dubai it was a university that built their satellite. This is why the contribution of universities is necessary and indispensable in the new development policies of nations. So we don't do personal research just to dip into the public purse for the sake of belly worship.

But to become part of the history of rare and free men, thanks to the power of mathematical thought reflected on the screen of space-time. To succeed in inscribing one's name in the annals of universal history is the dream of men who wake up early to have birds around them as their interlocutors. Africa's future rests on the shoulders of its best sons who always think big to do big for Africa and by Africa in the culture of the search for excellence in the skills of rare men.

Research into Light Waves through the lens of Quantum and Digital Poetry takes us on a great journey of discovery of the twelve-dimensional universe, through the great intelligence of our intellectual capacity to observe the light and its shadow, the image and its reflection, through the reality of symmetry, to re-establish the interconnection between the wave dimension and the physical dimension, through the poetry of Scientific Imagery.

The principles of the twelve-dimensional universe lay the profoundly scientific foundations of this new theory, providing a perfect link between two systems (wave and physics) that complement each other perfectly.

This Theory of the Twelve-Dimensional Universe is one of the great discoveries of the millennium, crossing philosophy, literature, classical and quantum physics, mathematics, especially mathematical logic, and spirituality.

A Theory of Everything. A grail turned towards the sacred feminine to sound the new genesis of man in innovation, creativity and spirituality leading to the development of nations on the basis of research and discovery in an original and authentic way without the slightest plagiarism on what exists. Even though we know that new things are built on what already exists. We must always go beyond what exists to bring new originality as a contribution to the progress of humanity without colour boundaries.

For any new theory, the coherence of the scientific approach is sufficient to grant it validity. It is now up to mankind to confirm or refute it with an example drawn from practical experience, based on the architecture of true truth that reflects natural analogies. A palpable example, also supported by Professor Paul YOTCHO, is the study of probability, which remains coherent in its approach scientific mathematical but practical experience has always shown that these probabilistic laws are not always well-founded. However, misguided people cannot question a theory without having the slightest intellectual capacity to understand it first and to see its limits in time and space through practical experience drawn from experimental soil on the basis of OCCAM's razor principle.

University civil servants (teachers) are not always capable of understanding a theory newly constructed in an interdisciplinary, multidisciplinary and transdisciplinary way, which is essential for access to true knowledge in the intellectual capacity to understand things in the light of the wisdom of common sense.

Professor Hassan BAKHSISS believes that the most daring theories in science are quantum space-time, parallel universes, string theory and elementary particles: the Higgs boson, neutrinos, the standard model of particles, etc.So there is a difference between a teacher and a professional researcher. Teaching at university doesn't make you a researcher or a scientist, still less one who has a monopoly on knowledge or even the intellectual capacity to understand all new knowledge in self-education. The researcher is the man of discovery, in constant dialogue with nature on the basis of intuition and mathematical astuteness, through a refinement of thought that is reflected on the screen of space-time. To receive the baptism of research is to be on the way to the intellectual capacity to fuse moral spirituality with exoteric spirituality to form a single entity in dimension twelve, the seat of the fusion and unification of things in the universe of digital derivation and integration in space.

It is the exoteric side of science (spirituality) that naturally provides the mysteries of research in discovery. So the science or research is not the prerogative of ordinary men. The word of an initiate, the honey of the chosen ones from the deepest bowels of the stars.

Our senses are not an exact science; they constantly lead us into illusion. Great researchers such as Einstein, Schrödinger, Bohr, Planck, Broglie and many others have revolutionised our perception of reality. The arrow of time that only goes from past to present to future is just an illusion; the fundamental equations of physics do not rule out its reversibility.Scientific research is a matter of realities, of vision and orientation, of plan and strategy, not forgetting the achievements and contributions on the basis of which each of us will be judged by the future in the universal light of reason. Each century is like a human being that needs to be fed. The century can only be nourished by innovation, invention, creativity and spirituality. This is why the scientific oath forbids us to live randomly in a century without finding something to nourish it. Researchers are always creators. He is an artist of the universe in his know-how, which he naturally accompanies with his knowledge of how to tame nature with the cries of eternity through a refinement of mathematical thought that is reflected on the screen of space-time.

CHAPTER I

WHAT IS THE TWELVE-DIMENSIONAL UNIVERSE THEORY?

As we try to understand one theory of the universe, another appears to change the course of history with new revelations. This means that we are still a long way from understanding and knowing all the secrets and mysteries of the intelligent universe.

The theory of the twelve-dimensional universe comes naturally from our intellectual capacity to observe light coupled with its shadow, the image and its reflection through the reality of symmetry drawn from practical experience on the architecture of true truth through a refinement of mathematical thought that is reflected on the screen of space-time.

This new theory, known as the twelve-dimensional universe, is a great observation, intelligible from a silhouette in the shadow of the southern light. From this I understood, through experience, that it is the shadow that makes the light exist. So the light that is coupled to its shadow is a function of an intelligence of fusion and unification of things in the universe to find their equilibrium from the image and its reflection by the reality of symmetry.

It is the Ondulatory and Physical dimensions that establish a perfect junction in the interconnection between light and its shadow; thus the image and its reflection through the reality of symmetry to take rehabilitate in the theorem of perfection where any positive integer that adds as many times as itself equals its second power. For example: let's add 3 as many times as itself. We have 3+3+3. Now $3+3+3= 3^2$. Hence $3^2 =3X3=9$. So 9 is the second power of 3. So 3 is a perfect square. And the relationship between the image and its reflection is perfectly established in this mathematical demonstration by the reality of symmetry. So, each of us is the reflection of the image that we are in the world beyond, to be rehabilitated in this sensible world by the the image and its reflection through the reality of symmetry. Now, what makes man is the real and the virtual merging to form a single entity. This is what leads me to demonstrate mathematically that man is a universe contained within himself that he discovers when he tries to understand reality on the basis of the theorem of perfection through the reality of symmetry. Man is a symmetry between the real and the virtual in dimension twelve.

Research into Light Waves through the prism of Digital and Quantum Poetry leads us to the discovery of the larger, twelve-dimensional universe that stretches from the infinitely small to the infinitely large, the seat of the fusion and unification of things in the universe of digital derivation and integration in space.

The particularity of this twelve-dimensional space-time is that time is a constant in one direction or another and linked to space. All temporal intelligence is a function of spatial intelligence.

So time is contained in space, expressed mathematically by the following relationship: d=ct; i.e. Cxt. Hence celerity multiplied by time creates a space of fusion by the technique of the surveyor in the temporal spanning of strings or landmarks.

Now, generally speaking, the twelve-dimensional universe is the temporal straddling of century markers in the concept of point and axial symmetry compared to Minkowski's universe by the light of the mind, which eliminates time in a mathematical demonstration to compose solely with numbers and space.In other words, a dual system of real and virtual reference points in relation to an illuminated imaginary mirror, using a style of derivation and digital integration in space. In other words, the bringing together of shadow and light, and the fusion of the light of spirit and matter in a creative movement.

Consequently, Special Relativity and General Relativity are contained within the larger and wider twelve-dimensional space-time, the seat of fusion and unification to re-establish the equilibrium of the universe from the infinitely small to the infinitely large. And in relation to the earth.

This twelve-dimensional space-time is the simplicity and complexity of the world on the one hand, and the unity and diversity of the world on the other. where all creation is musicality, a circular survey of the universe with the earth rotating on itself.

It is a space for the correlation between Science, Spirituality and Art in Research and Development, the aim of which is to link science to spirituality and art on the basis of reason contained in emotion. For all rational intelligence is a function of emotional intelligence. Hence the reason contained in emotion is the seat of science in practical experience.

Twelve-dimensional space-time is the universe of the re-establishment of the marriage of innovation and the creation of the spirit in spirituality that leads us to the productivity of new knowledge, essential to the growth of African nations in inventive and creative genius in mathematical relation to real need. It is a fusion of two cones of light and darkness, the aim of which is to re-establish the link between science, spirituality and art in research and development. This space also requires a fusion in a square-based pyramid whose sides are triangular rectangles whose rotation in one direction generates a transcendent and ascending circle in the light that couples with its shadow, the image and its reflection through the reality of symmetry.

Twelve-dimensional space-time, then, is the crossing of reference points into a new frame of reference where body and mind merge into four transcendent points, called horizons or the point of intersection of the circle and the inscribed square.

Twelve-dimensional space-time links science with spirituality and art, whose words are algorithms containing energy, which creates the matter of the void. To exist in twelve-dimensional space-time is to be outside oneself towards someone other than oneself.

The theory of the twelve-dimensional universe takes us out of the physical system and into a wave system. This invention is a part of me drawn from the experimental soil on the basis of OCCAM's razor principle that I am making a wonderful gift to humanity as it progresses across the screen of mathematical space. A theory of everything, a grail turned towards the sacred feminine to fertilise the millennium through innovation, invention, creativity and spirituality, leading us to the growth of the twelve dimension of a strong economy of defence and security for Africa's continental sovereignty.

CHAPTER II

THE FOUNDATIONS OF THE TWELVE-DIMENSIONAL UNIVERSE THEORY

Twelve-dimensional space-time is a universe whose foundations are drawn from the experimental soil of anthropological, philosophical, cognitive and exact sciences...

The bricks that form the backbone of the Theory of the Twelve-Dimensional Universe are alloys derived from the creative transdisciplinary excursions of the soul of the civilisations of the black peoples in their most accomplished versions from ancient times to the present day.

The foundations of twelve-dimensional space-time are above all spiritual, scientific and artistic, stemming from research into the Light Wave through the lenses of Digital and Quantum Poetry in the first fundamental relationship of surveying, L=l+d.

Based on OCCAM's razor principle, we have the following principles:

-Everything that appears to the human mind is the expression of a reality;

-All reality is contained within a precise space-time;

Four-dimensional physical space-time is contained within six-dimensional space-time;

-All space-time is the seat of energy, memory and intelligence;

-All beings are governed by universal scientific laws;

-All beings are controlled dynamic systems;

-All rational intelligence is a function of emotional intelligence;

Six-dimensional space-time is contained within twelve-dimensional space-time.

These principles are in fact fundamental theorems, propositions that must be scientifically proven.

CHAPTER III

THE FUNDAMENTAL THEOREMS OF THE TWELVE-DIMENSIONAL UNIVERSE THEORY

For a better understanding of our study, we intend to provide evidence for the scientific assertions that form the architecture of our Theory of the Twelve-Dimensional Universe.

First fundamental theorem

I- Everything that appears to the human mind is the expression of a reality.

Demonstration:

We know that :

Everything that appears to the mind is thought.

1- The brain is the seat of thought, materialised by electrochemical signals.

2- The brain has specialised areas and/or long- and short-term memories for receiving and processing images, sounds, emotions, smells and other information.

3- All conscious or unconscious cerebral information (dreams, thoughts, mental exercise, etc.) is measurable or observable (recorded by cerebral imaging techniques such as Positron Emission Tomography or nuclear magnetic resonance).

4- Everything that appears in the mind is recorded by the brain in the form of electrochemical signals that can be detected using brain imaging techniques.

Facts: 1, 2, 3 and 4. Now

5- All this that is detectable is measurable and objectively observable reflects a physical reality.

So

6- Everything that appears to the human mind is the expression of a reality

Second fundamental theorem

I- All reality is contained within a specific space-time.

Demonstration:

From an epistemological reading of the First Fundamental Theorem, it can be seen that :

1- A reality is a universe or space randomly presented to the mind in a given time interval.

A spatiotemporal structuring of the vectors of four known forces in the universe (gravitational interaction, electromagnetic interaction, weak interaction and strong interaction) shows that :

2- Any particle associated with a force field lives in an extremely precise space-time

On the other to observation of clocks (Atomic, Biological, Quantum) we notice that :

3- There is no time without movement.

4- There can be no regular movement of an object without a force acting on it.

5- There is no force without precise laws, known or unknown, without precise space-time.

6- The stability of universes depends on the space-time pressure of natural phenomena.

7- All reality is not contained in a precise space-time.

So

The propositions: 6.5, 4, 3 and 2 arefalse. Which confirms the demonstration.

Third fundamental theorem

Four-dimensional space-time is contained within six-dimensional space-time.

Demonstration:

We know that the four-dimensional space-time **E44 is** expressed mathematically by the following relationship.

1) $E44 = \{x, y, z, iict\}$

X, Y, Z being spatial coordinates and $iict$ the time coordinate, where $c =$ 300,000,000 m/s represents the speed of light in a vacuum; $i = \sqrt{-1}$ a pure imaginary and t time.

Gold

The light is propagates generally in space-time à three-dimensional space-time.

$E33 = \{xx, yy, zz\}$. Velocity vector.

The celerity can be broken down into three components: Cx; Cy; Cz

Hence the formula

2) $C = Cx + Cy + \boldsymbol{Cz}$. Vector sum of the above components. **Hence**

3) $ICt = i(C_x + C_y + C_z)t$

Expression 3. Shows that the fourth coordinate ICt is a vector quantity that can be decomposed into three imaginary components:

$iicxt$; $iicy\ t$; $iiczt$.

Consequently, the four-dimensional space-time E_{44} is contained within a six-dimensional space-time. In other words

4)$_{66}$ $E = \{xx,\ yy,\ zz,\ iic_x\ t,\ iic_y\ t,\ iic_z\ t\}$This which confirms the demonstration.

Fourth fundamental theorem

I- All space-time is home of energy, of memory and intelligence.

Breaking down this complex proposition into three components or simple propositions.

1- All space-time is the seat of energy;

2- All space-time is the seat of memory;

3- All space-time is the seat of intelligence.

Demonstration 1 :

All space-time is the seat of energy.

We know from experiments in quantum physics that :

1- The vacuum contains energy (vacuum matter can be created, CASIMIR effect)

2- There is a minimum energy in any physical system.

Hence

$E0 = 6.626.\ 10^{-34}\ j$

Gold

3- All space-time (considered as extended without matter) is a physical system.

So

4- All space-time is the seat of energy.

This non-empty space-time is called energy space-time.

Demonstration 2 :

We know that :

1- All space-timeis space- time energy (see demonstration 1).

2- Energy can be transformed into matter and matter into energy in accordance with Einstein's formula:

$E = MC^2$ This gives the equivalence relation.

3- Space-time energy = Space-time matter

Gold

4- All matter obeys one or more laws of physics.

5- There is no obedience without substance.

So

6- All space-time-energy or space-time-matter has a memory.

7- Space-time-energy-memory is the concept that expresses this reality.

Demonstration 3 :

All space-time is the seat of intelligence.

It is perfectly obvious that :

1- There is no intelligence without algorithms.

2- There is no algorithm without calculation.

3- There is no calculation without memory.

4- There is no calculation without law.

5- There are no laws without language.

6- There is no language without syntax.

7- There is no syntax without semantics.

8- There is no semantics without intelligence.

If all space-time is not the seat of intelligence. Then

9- There are no semantics, syntax, language, calculation or memory in any universe. Which is contradictory.

So

All space-time is the seat of energy, memory and intelligence. This confirms the demonstration.

Fifth fundamental theorem

I- All beings are subject to universal scientific laws, both known and unknown.

Demonstration:

Modern physics teaches us that :

1- The universe is subject à four laws or forces.

The existence of matter is linked to the balance of these four forces:

2- Gravitational interaction ensures the stability of celestial bodies.

3- Electromagnetic interaction stabilises electrons around atomic nuclei.

4- The weak interaction controls radio activity.

5- The strong interaction ensures the cohesion of protons and neutrons in the nuclei of atoms.

Gold
6- A kingdom can be mineral, vegetable, animal or symbolic.

So
7- All beings in the visible or invisible universe are subject to known or unknown laws.
8- The existence of dark matter and the mysterious energy that makes up nearly 95% of the mass of the universe are subject to unknown laws.

We also know that :
9- All information systems are subject to universal scientific laws of organisation, both known and unknown.

Gold
All mineral, biological, social and symbolic beings, Mental or organic ar are information systems.

So
All beings are subject to universal scientific laws. This confirms the demonstration.

Sixth fundamental theorem

I- All beings are controlled systems.
Demonstration:

Assume that :
1- Not all beings are controlled systems.

We know that :
2- Every controlled system has a data and programme memory.

3- A controlled system is a feedback loop.

4- A programme is an ordered sequence of instructions or laws.

5- Elementary particles, atoms and molecules obey unknown universal laws of physics.

6- Dark or invisible matter or mysterious dark energy obeys unknown laws of physics.

7- Everything subject to the law is ordered.

8- All beings are dynamic systems subject to the laws of organisation.

If the first proposal is correct.

So..,

The facts reflected in the proposals :

8, 7, 6, 5, 4, 3, 2, 1, are false. Which is contradictory.

CHAPTER IV

THE COROLLARIES OF THE THEORY OF THE TWELVE-DIMENSIONAL UNIVERSE

For a better understanding of our study, we believe that we can contribute propositions immediately derived from another, which contribute to the construction of our Theory of the Twelve-Dimensional Universe.

First fundamental corollary

I- All spirituality is rooted in humanity.

Demonstration:

We know that :

All spirituality is in the realm of thought.

1- Science is a refinement of ordinary thought that embodies a spirituality of peoples.

Gold

2- There is no science, no humanity, without spirituality.

3- There can be no science without poetry, which reflects the spirituality of peoples.

4- All spirituality is the expression of the soul of a civilisation.

So

5- All spirituality is root of humanity. This which confirms the demonstration.

Second fundamental corollary

I- All transformations in the animal, mineral, plant and symbolic kingdoms obey known or unknown universal physical laws.

Demonstration:

Assume that :
Not all transformations in the animal, mineral, vegetable and symbolic kingdoms obey known or unknown universal physical laws.

We know that :
1- Every transformation has a data memory with an operating system.
2- A transformation is an ordered information system that obeys known or unknown universal laws of physics.
3- Everything subject to the law is ordered.

4- Everything in the animal, mineral, plant or symbolic kingdoms is a dynamic system subject to laws.
If these propositions or corollaries are false.
So

Which is contradictory.

Third fundamental corollary

I- All totemic power is the expression of a dynamic transformation subject to known or unknown physical laws.

Demonstration:
We know that :
Every totemic power is a transformation that has a controlled system of

exploitation, whether in the animal, mineral, vegetable or symbolic kingdoms.

1- All totems are physical transformations governed by known or unknown universal laws.
2- A totem is an energy with a memory and an intelligence for controlled transformation.

Gold
3- Totems are beings from the animal, mineral, plant and symbolic kingdoms that are subject to known and unknown universal scientific laws.
So

All totemic power is the expression of a dynamic transformation subject to known or unknown physical laws. This confirms the demonstration.

CHAPTER V

CONSTRUCTION OF THE THEORY OF THE TWELVE-DIMENSIONAL UNIVERSE THROUGH LIGHT AND SHADOW, THE IMAGE AND ITS REFLECTION THROUGH THE REALITY OF SYMMETRY

We know that twelve-dimensional space-time comes from light, which is naturally coupled to its shadow, the image and its reflection through the reality of symmetry. This so-called twelve-dimensional universe plunges us into a physical system to approach another so-called wave system in the temporal straddling of landmarks or strings using the surveyor's technique to rehabilitate in the theorem of perfection where any positive integer that adds up as many times as itself equals its second power. Twelve-dimensional space-time is the seat of the fusion and unification of things in the universe of digital derivation and integration in space. This twelve-dimensional universe is a reality derived from practical experience based on the great intelligible observation of light, which is coupled to its shadow, image and reflection through the reality of symmetry.And naturally extends from the universe of the infinitely small to the infinitely large in the temporal straddling of century markers in the concept of point and axial symmetry, compared to Minkowski's universe by the light of the mind that eliminates time in a mathematical demonstration to compose solely with numbers and space. In other words, a dual system of real and virtual reference points in relation to an illuminated imaginary mirror, using a style of derivation and digital integration in space. In other words, the bringing together of shadow and light and the fusion of the light of spirit and matter in a creative movement.It is also the fusion of the rational universe and the emotional universe to form a single entity. Reason is contained within emotion, the seat of the awareness and of intelligence in practical experience.

All rational intelligence is a function of emotional intelligence. Let's build twelve-dimensional space-time through the temporal spanning of the reference points or strings of special and then general relativity, the fundamental relationship of surveying and the theorem of perfection through light coupled to its shadow, the image and its reflection through the reality of symmetry.I- Four-dimensional space-time (special relativity) is contained within six-dimensional space-time (general relativity).

Demonstration:

We know that the four-dimensional space-time **E44 is** expressed mathematically by the following relationship.

5)$E44 = \{x, y, z, iict\}$

X, Y, Z being spatial coordinates and $iict$ the time coordinate, where c = 300,000,000 m/s represents the speed of light in a vacuum ;

$i = \sqrt{-1}$ a pure imaginary and t time. Now

The light is propagates generally in space-time à three-dimensional space-time.

$E33 = \{xx, yy, zz\}$. Velocity vector.

The celerity can be broken down into three components: $Cx; Cy; Cz$

Hence the formula

6)$C = Cx + Cy + \boldsymbol{Cz}$. Vector sum of the above components.

Hence

7)$ICt = i(C_x + C_y + C_z)t$

Expression 3. Shows that the fourth coordinate ICt is a vector quantity that can be decomposed into three imaginary components:

$iicxt; iicy\, t; iiczt$.

Consequently, the four-dimensional space-time E_{44} is contained within a six-dimensional space-time. In other words

8) $_{66}E = \{xx, yy, zz, iic_x t, iic_y t, iic_z t\}$

This confirms the demonstration.

Six-dimensional space-time is contained within twelve-dimensional space-time by the light that is coupled to its shadow, the image and its reflection by the reality of symmetry.

Here's how it works.

Any positive integer that adds up as many times as itself equals its second power. This is the theorem of perfection. In other words, the light and its shadow, the image and its reflection through the reality of symmetry.Now, man is a symmetry between the real and the virtual, merging to form a single entity. That's why, in the world beyond everything is perfect, the square is perfect, the horse is perfect. So the sensible world is just an approximate photocopy of the world beyond in physical experience, to be rehabilitated in the theorem of perfection where any positive integer that adds up as many times as itself equals its second power.Hence each of us is the reflection of the image that we are in the world beyond through the light that couples with its shadow through the reality of symmetry.

Every silhouette is a function of light. It is a part of light that generally propagates in three-dimensional space-time. So light needs shadow to exist. Because there is no light without shadow, and no shadow without light. Shadow is the fuel of light in mathematical relation to the real and the virtual. That is to say, reflections or counter-reflections; counter-reflections or reflections of digital derivation and integration in space. It is the fusion of things to form a single entity between the material and the immaterial in the concept of point and axial symmetry.

Cône lumineux			*Cône ténébreux*
33			33′
0	y		0 y'
x			x'
$x, y, z, iict$ $iicxt, iicyt, iiczt$ x, y, z, XX, Y, Z			$x', y', z', iic't$ $iic'xt, iic'yt, iic'zt$ x', y', z', XX', Y', Z'
		Fusion des répères réel et virtuel	

$E12 = x, y, z, iicxt, iicyt\ iiczt, x', y', z', iic'xt, iic'yt, iic'zt.$

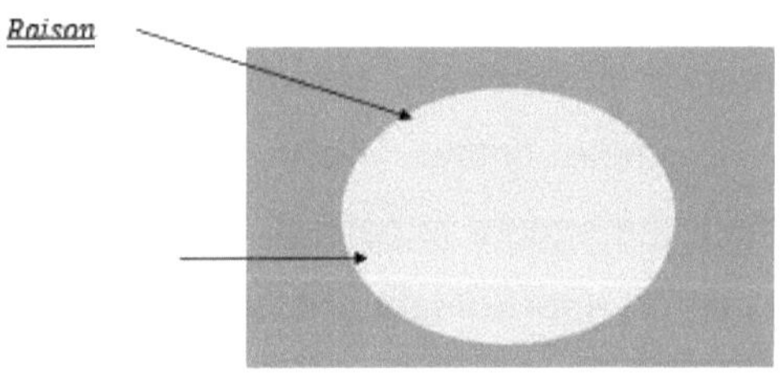

Fusion des univers rationnel et émotionel

$R \subset E$

CHAPTER VI

THE CORRELATION BETWEEN SCIENCE, SPIRITUALITY AND ART IN RESEARCH AND DEVELOPMENT

The mission of twelve-dimensional space-time is to lead us towards the construction of a new world of the economy of thought, time and space in the correlation between science, spirituality and art in research and development.The aim of twelve-dimensional space-time is to link science, spirituality and art so that they form a single whole in a single larger universe, extending from the infinitely small to the infinitely large in the measurement of distance and direction, the speed of light in a vacuum, the fixing of energies and the positioning of particles in the space-time continuum.The world of the economy of thought, time and space is that of the poetisation and spiritualisation of man's behaviour towards the sacred through a refinement of mathematical thought in a dual system with the two cones of light and darkness, to be rehabilitated in the theorem of perfection.Where any positive integer that adds up to so many times that itself equals its second power. This is where man becomes even more spiritual and altruistic, always thinking well and justly for others on the screen of mathematical space.Twelve-dimensional space-time is the fusion between science and art in spirituality, the aim of which is to teach us reason in the elegance of art that reflects our observations and feelings contained in the DNA of our socio-cultural environment; where we are condemned to make contributions to the progression of humanity.Twelve-dimensional space-time also means reorganising the world of research and development into a new frame of reference. the transmission of knowledge from the experimental ground on the architecture of truth, which is a major problem in Africa.Spirituality is the key to Research and Innovation leading to development based on productivity and creativity in culture and art. Spirituality always manifests itself in the culture of a people and is expressed in art to trigger inventive and creative genius in mathematical relation to real need.All scientific, technical, technological and even artistic inventions are naturally a result of spirituality in practical experience drawn from experimental soil. So all artistic creation is a spiritual representation materialised in real need by inventive and creative geniuses.

Spirituality is the light of the spirit, or even a spiritual plane that manifests itself in us with images, which we have to capture by means of a light wave that is zebra-striped and refracted on the basis of intuition and mathematical astuteness.

The role of intuition is to reveal to us the nature of the images captured by the light wave during the manifestation of a spiritual plane within us. Mathematical astuteness, on the other hand, is the materialisation in real time of the images captured by the light wave and then revealed by intuition in a permanent dialogue between nature and man. So mathematical astuteness is inventive and creative genius.

Consequently, the zebra and refracted light wave is a cosmic image-capturing energy that manifests itself in us through a refinement of thought based on intuition and mathematical astuteness to re-establish the link between nature and man through the channel of spirituality or the spiritual plane. Hence Scientific Research is the intellectual capacity to understand, to question and to make nature react so that it reveals itself to us on the basis of intuition and mathematical astuteness in a permanent dialogue between nature and man. Spirituality is the only bridge that can exist between science and man to faithfully imitate nature in its authentic configuration that reflects natural analogies on the architecture of truth. No concept can be accurately defined without the slightest dialogue with nature drawn from experimental soil. It is through experience that man succeeds in discovering the truth, whose error is a tool of permanent learning that naturally leads us to the school of knowledge where we are all learners in the void of nature, in the step of differential varieties with fractal dimensions.Twelve-dimensional space-time helps us to understand the profound meaning of the universe and the correlation between science, spirituality and art in research and development. And to determine mathematically its link with the space that surrounds us, that carries us. My scientific quest, applied to Digital and Quantum Poetry in a twelve-dimensional space-time, is all about rediscovering our ancestors, recognising the path taken by African humanity and giving them back their power over our modern lives.

Poetry is not a song sung from within with instruments of the heart. Rather, it is the emotions that speak to us in a silent form of writing that kills language with the step of differential varieties.

Poetry is the source from which springs the scientific impulse to philosophise in order to understand the book of nature written in the language of the fractal mathematics of pure nature. In other words, a primitive of the function derived from science in the memory of the universe. Poetry is the matrix of all the sciences linked to spirituality and art, which must fertilise the millennium with innovation in research and development based on productivity and creativity.

You can't do research or science if you're not in league with poetry as the foundation of emotion, the seat of consciousness and intelligence; of our thoughts and ideas; of our feelings and observations; of our actions and impressions... which the brain carries out in thinking action. Poetry is also the true seat of science in the history of African cultural thought in the deep Africa of the pyramids. The great scientists are always great poets, naturally immersed in a rational and emotional universe in fusion. to become one in the unification of things from derivation to integration.So poetry is a rational and emotional intelligence within us to understand the universe and its language in forms and numbers from the primitive to the derivative or from the derivative to the primitive and then integration into a larger and wider space-time that extends from the infinitely small to the infinitely large. Poetry is the tradition of the initiated, the guardian of a people's culture. Tradition is the very essence of science by the light of the mind. It is our historical and cultural identity in the history of African cultural thought. Poetry is the science of the initiated in its spiritual dimension, the queen of sciences and its servant, the architecture of reality in practical experience.So writing a collection of poems is not the same as writing poetry or being a poet. But it is to throw a cry from eternity onto a page in order to rebuild oneself in the foundation of the memory of writing. Hence the Poetry of Scientific Imagery in a twelve-dimensional space-time is a fusion of science with spirituality and African art to form one in interdisciplinarity and innovation. Where reason is contained within emotion. The universe began with, and electromagnetics was. Our poetic statements need to be reformulated by scientific disciplines. The mind is the centre of vibratory suspension, where a different way of thinking can be born, a different vision of the world that seems to be a vision of another world.Poetry has the capacity to trigger an open link within what is, in which everything is different from what it usually is, and at the same time reveals itself as what is in a space-time of fusion and unification. Poetry seeks to embody the non-teleological universe that science hints at. Digital and quantum poetry explores the states within which we are propelled, transformed and sublimated into a new frame of reference; where the poet of noon puts forward a (pre-) observation that neither claims to encompass what he has perceived, nor to give it a specific meaning. separate existence outside the morpho-poetic field of an entanglement of language and the solar world.

Poetry does not fill a vacuum of meaning; it places things in a vacuum that shows the relationships between them. Poetry creates forms in which we experience ourselves and our environment in the universe through a refinement of mathematical thinking.

Digital and quantum poetry is the basis of all language. Because a language is a poem in perpetual metamorphosis from shadow to light, balanced by a silhouette in the shadow of midday merging with the light of the spirit. The aim of digital and quantum poetry in a twelve-dimensional space-time is to materialise ideas and give them form in a mathematical relationship with real need through a refinement of thought.

My poetry creates a universal language of algebraic rigour, free from images and the spatial perception of all human experience in the space-time of the correlation between science, spirituality and art in research and development.

The poetry of scientific imagery is the marriage of science, spirituality and art, which has the capacity to go beyond our representations and beyond our intuition and mathematical astuteness to faithfully imitate nature in its authentic configuration that reflects natural analogies. It is the marriage of algebra, analysis and geometry that has the privilege of developing thought beyond our ability to visualise things. The poetry of scientific imagery in a twelve-dimensional space-time is descriptive geometry where the algebraic expression $a + b = c^{222}$, the second fundamental relationship of surveying to develop immaterial forms and numbers to speak of reality. So there are equations with dimensions or images that have a performative value. What the mathematician is to the material world, the poet is to the intelligence that runs through the world of ideas.

The poet is the one who has the ambition to reach life in the totality of its becoming to live in the infinite time where lives are micro vortices as cosmic energy that swirls and circulates within us in differential varieties. So the poet is a whirlwind who is always penetrating the secrets and mysteries of the intelligent universe.

In physics, the particle is described by a lower wave function (y(r)), which cannot be visualised. This wave function is a sum of information expressed by a complex set of variables in terms of probability. In the poetry of scientific imagery in a twelve-dimensional space-time, the particle of meaning is neither a word nor an image, but rather a phonic gesture that tends towards becoming, an expectation that opens onto the globality of relationships. Quantum behaviour can only be observed on isolated particles and cannot be transported on a human scale. All things are intelligible and sensitive. The poetry of scientific imagery in a twelve-dimensional space-time recognises an expansive and living intelligence in the correlation between science, spirituality and art in research and development. Some of the difficulties of particle physics can or must be answered by the poetry of scientific imagery in a twelve-dimensional space-time where I am the father-inventor in the spirituality of the Mibaambas-Mikwissis peoples, the guardian of culture in art.The understanding of the universe and its interpretation through poetry in a twelve-dimensional space-time is a great scientific invention of the millennium by Audrey Kibamba de Bouansa, a station of light in the correlation between science, spirituality and art in research and development.

CHAPTER VII

APPLICATIONS OF THE SPACE-TIME CORRELATION BETWEEN SCIENCE, SPIRITUALITY AND ART IN RESEARCH AND DEVELOPMENT

Any theory of the physical universe built on scientific, artistic and spiritual foundations drawn from experimental soil has applications in all fields (interdisciplinary and multidisciplinary) that are rooted in the mysteries of life; in the totality of knowledge and in the knowledge of the totality. Twelve-dimensional space-time is a new frame of reference that is leading us to reorganise the world of work in the digital and quantum economy of thought, time and space.This Kibambist universe is a great revolution, first and foremost cultural, then scientific, technical and technological, artistic and spiritual, based on interdisciplinarity and innovation to find solutions to global problems through productivity and creativity leading to development. Twelve-dimensional space-time is capable of revolutionising the world of quantum computing, with the construction of new, ultra-powerful quantum computers that save thought, time and space by merging and unifying digital derivation and integration in space. Twelve-dimensional space-time can also be applied to the world of robotics, with the production of quantum sapiens robots capable of organising the world of the economy of thought, time and space through interdisciplinarity and innovation.In Africa, this twelve-dimensional space-time is the key to the African renaissance in research and innovation based on productivity, creativity and spirituality in an African education that takes account of African realities. So it's a question of the realities of vision and direction, of plan and strategy, not forgetting the contributions to the progress of humanity on the screen of mathematical space. This invention values the inventive and creative genius of Africa in the cultural, scientific, artistic and spiritual spheres, where we are called upon to make original contributions in order to make a fresh start and enter the history of mankind with the weapons of intelligence.

Twelve-dimensional space-time is a revolution in quantum or particle physics at the dawn of the third millennium in interdisciplinarity and innovation on the basis of the Poetry of Scientific Imagery capable of responding favourably and clearly to certain concerns of particle physics in twelve-dimensional space-time. The theory of the twelve-dimensional universe also leads us to rediscover the equilibrium of our world through the disorder that is coupled to its order through the reality of space-time. It is disorder that brings order into existence. Hence order is an integral part of disorder, in order to rediscover the balance of things in the universe.

In every universe, things are linked to each other in the foliage of time. And in relation to the moving earth of surveying. Disorder, understood as space and order, is time linked to space, and which remains constant in one direction or another between two systems (physical and wave) to operate a perfect junction. The Theory of the Twelve-Dimensional Universe is the centre of the interconnection between disorder and order, to restore balance to our world through the reality of time linked to space to form one. And to rehabilitate in the theorem of perfection where any positive integer that adds up as many times as itself equals its second power. So order and disorder are one. Consequently, they are inseparable if we are to regain equilibrium.

CONCLUSION

Our inability to understand and define concepts properly makes us illiterate or illiterate people who have been to school but do not understand school as others do in the universal light of reason. Universities train students, while society trains people through practical experience.The one who knows is the one who wakes up very early to have birds around him to talk to. Because he says what others don't understand. And he does not see things on the same wavelength as his fellow men. The imagination of the scientist is that of a small child of six to eight years old who sees things, sometimes terrifying, in the world to which he is linked that others cannot see or understand as he does.

The knower is an enlightened and rare spirit who knows only how to love, seek and search for the true truth in the culture of the search for excellence in human competence. They know not how to envy, hate, envy, let alone slander. Knowledgeable people are as rare as an eclipse, and as hard to find as gold and diamonds. So you have to be a scholar yourself before you can call someone a scholar on the basis of their achievements and contributions in the fields of science and literature.

Our existence is essentially mathematical as the roots of humanity. Life began with mathematics and in mathematics to build the world with words as the instruments of literature and science. Without words there is no freedom to think, to reflect or to exist. Words are an expression of divine life. Human thought is mathematical. And everything is done with thought by thought.Nothing great can be achieved without some mastery of mathematics. The great discoveries that have revolutionised human thought are rooted in mathematics. The human being is the image of the universe. Its appearance at the atomic level is in reality and in part filled with emptiness (space at the atomic level between nucleus and electrons, between atoms) while the atoms that make us up are governed in relation to each other by strong attraction. From then on, we can legitimately consider that our body contains several levels of dimensions (at the level of the infinitely small elements inside the atomic nucleus, then the molecular nucleus, etc.) that the atoms and molecules that make us up twist space-time at different levels, just like more or less dense bodies here or there in the universe.

Proof of this can be found in Lightwave Research, which shows that the human body modifies the trajectory of particles travelling close to it. As thought is organised by the arrangement of atoms and molecules, why not also consider that thought can be a lever that further influences and twists space-time at our level and at different levels of our body.

The earth, being the element of matter, is a sphere that floats in the universe. And that in the universe there is no up and down. In other words, north and south do not exist in the universe. So it's not insignificant that Europe can be represented at the north of a world map and Africa at the bottom. This has no mathematical significance in the universe. Just as Africa can be represented at the north of a world map and Europe at the bottom. The universe contains itself in its entirety, with no borders or edges. To look up at the night sky is to gaze into infinity - its dimensions incomprehensible and therefore meaningless. Our passion for learning is our survival tool. In every universe there are N dimensions. And not all the dimensions of the universe have the same meaning or the same interpretation.Leave it to the future to tell the truth and to evaluate everyone according to their achievements or contributions to the progression of humanity on the screen of mathematical space. The present is theirs, the future, for which I have really worked, is mine.Intelligence is not the ability to store information, but knowing where to find it. The information is always in the memory of the intelligent universe, which needs to be tamed. by obeying it. The universe is a poetry of science written in the language of nature's pure fractal mathematics.

SCIENTIFIC PUBLICATIONS

✓ Comment je vois les Belles Lettres: Théorie de la Poésie de l'Imagerie Scientifique dans un espace-temps à douze dimensions, published by Le Lys Bleu Paris France in 2021. Author Audrey Kibamba de Bouansa gare de lumière.

✓ La Théorie de l'univers à douze dimensions sur l'écran de l'espace-temps, published by European University Publishing in Düsseldorf, Germany, in 2021. Author Audrey Kibamba de Bouansa gare de lumière.

✓ Nouvelles Politiques de Développement des Nations Africaines dans un espace-temps à douze dimensions, published by European University Publishing in Düsseldorf, Germany, in 2021. Author Audrey Kibamba de Bouansa gare de lumière.

✓ L'Afrique, Mémoire de l'Humanitude whose Essence and Urgency are the struggle of our time in a twelve-dimensional space-time, published by European University Publishing in Düsseldorf, Germany, in 2021. Author Audrey Kibamba de Bouansa gare de lumière.

✓ Connaissance et Humanité: les vrais diplômes dans la production intellectuelle dans l'auto-éducation, published by European University Publishing in Germany in Düsseldorf in 2021. Author Audrey Kibamba de Bouansa gare de lumière.

✓ What is the Poetry of Mathematical Equations in the marriage of innovation and creation in a twelve-dimensional space-time, published by European University Publishing in Germany in Düsseldorf in 2021. Author Audrey Kibamba de Bouansa gare de lumière.

✓ Positioning the Spatio-Temporal Continuum in the First Fundamental Relationship of Surveying, published by European universities in Germany in Düsseldorf in 2021. Author Audrey Kibamba de Bouansa gare de lumière.

✓ Discourse on the Theory of Poetry of lyrical surveying in a twelve-dimensional space-time, published by European University Publishing in Germany in Düsseldorf in 2021. Author Audrey Kibamba de Bouansa gare de lumière.

✓ Les Douze Clés de la Connaissance sur l'Onde Lumineuse avec les lunettes de la Poésie Numérique et Quantique dans un Grand Carré, published by European University Publishing in Düsseldorf, Germany, in 2021. Author Audrey Kibamba de Bouansa gare de lumière.

✓ Initiation à l'Epistémologie Numérique et Quantique dans un espace-temps à douze dimensions, published by European University Publishing in Germany in Düsseldorf in 2022. Author Audrey Kibamba de Bouansa gare de lumière.

✓ Le Bloc Fédéral Panafricain Mécanismes des Réformes du Système Educatif Africain dans les Lumières des Humanités Classiques Africaines, published by European University Publishing in Germany in Düsseldorf in 2022. Author Audrey Kibamba de Bouansa gare de lumière.

✓ Denis Sassou N'Guesso at the heart of the challenges of deep African diplomacy in a multipolar world of balance and peace, published by European University Publishing in Germany in Düsseldorf in 2022. Author Audrey Kibamba de Bouansa gare de lumière.

✓ The Correlation between Science, Spirituality and Art in Research and Development,published by European University Publishing in Düsseldorf, Germany, in 2024. Author Audrey Kibamba de Bouansa gare de lumière.

✓ Introduction to the Theory of the Twelve-Dimensional Universe to restore balance to our world, published by Editions Universitaires Européennes in Düsseldorf, Germany, in 2024. Author Audrey Kibamba de Bouansa gare de lumière.

✓ L'arpentage lyrique, published by Muse in Düsseldorf, Germany, in 2021. Author Audrey Kibamba de Bouansa gare de lumière.
✓ L'aube des chants d'initiés, published by Le Lys Bleu Paris France in 2020. Author Audrey Kibamba de Bouansa gare de lumière.

CONFERENCE DEBATES

✓ On the theme of Quantum Thinking at the heart of innovation. Author Audrey Kibamba de Bouansa gare de lumière on Mobali Makassi's youtube channel on 19 February 2023 at 7pm Paris time France.

✓ On the theme of Digital and Quantum Epistemology, the key to Research and Development for African Nations. Author Audrey Kibamba de Bouansa gare de lumière on the youtube channel Causons d'Afrique on 17 November 2022 at 8pm Paris time France.

✓ On the theme Cleaning the Human Skeleton and Genome with Natural Light and Digital Pressure. Author Audrey Kibamba de Bouansa light station on the voice of the diaspora on 7 June 2022 at 3 p.m. Abidjan time in Côte d'Ivoire.

✓ On the theme How I see the Belles Lettres in a twelve-dimensional space-time. Author Audrey Kibamba de Bouansa gare de lumière on the youtube channel huit milles tambours d'Afrique on 20 March 2022 at 5 p.m. Brussels time in Belgium.

✓ On the theme of the Correlation between Science, Spirituality and Art in Research and Development, and the Theory of the Twelve-Dimensional Universe. Author Audrey Kibamba de Bouansa light station at the Faculty of Science and Technology of the Université Marien Ngouabi on 20 February 2024 at 10 a.m. Brazzaville time.

REFERENCES

1. The theory of the twelve-dimensional universe on the screen of space-time,Audrey Kibamba de Bouansa light station;

2. Comment je vois les belles lettres: théorie de la poésie de l'imagerie scientifique dans un espace temps à douze dimensions,Audrey Kibamba de Bouansa light station;

3. The positioning of the space-time continuum in the first fundamental relationship of surveying,Audrey Kibamba de Bouansa light station;

4. The correlation between science, spirituality and art in research and development,Audrey Kibamba de Bouansa light station.

Printed by Books on Demand GmbH, Norderstedt / Germany